# My Muscles

by Rachel Rose

Minneapolis, Minnesota

Credits: Cover, all background, © Piotr Urakau/Shutterstock; cover, 7, 23 © BlueRingMedia/Shutterstock; cover, 4, 9, 12, 16, 20, 22 muscle illustration © Shutterstock; 4, 8, 18, 20 Sergey Novikov/Shutterstock; 5 Samuel Borges Photography/Shutterstock; 4 TinnaPong/Shutterstock; 5 Robert Kneschke/Shutterstock; 6 LightField Studios/Shutterstock; 9 BlueRingMedia/Shutterstock, 9 boy VALUA STUDIO/Shutterstock; 10 Fotokostic/Shutterstock; 11 Youproduction/Shutterstock; 12 Hung Chung Chih/Shutterstock; 13 Krakenimages.com/Shutterstock; 14 Roman Samborskyi/Shutterstock; 15 (fist) Prostock-studio/Shutterstock; 15 (heart) Nixx Photography/Shutterstock; 17 Magic mine/Shutterstock; 19, 21 Pressmaster/Shutterstock.

President: Jen Jenson
Director of Product Development: Spencer Brinker
Senior Editor: Allison Juda
Associate Editor: Charly Haley
Designer: Oscar Norman

*Library of Congress Cataloging-in-Publication Data*

Names: Rose, Rachel, 1968- author.
Title: My muscles / by Rachel Rose.
Description: Fusion books. | Minneapolis, Minnesota : Bearport Publishing Company, [2022] | Series: What's inside me? | Includes index.
Identifiers: LCCN 2021045035 (print) | LCCN 2021045036 (ebook) | ISBN 9781636914442 (library binding) | ISBN 9781636914510 (paperback) | ISBN 9781636914589 (ebook)
Subjects: LCSH: Muscles--Juvenile literature.
Classification: LCC QP301 .R6544 2022 (print) | LCC QP301 (ebook) | DDC 612.7/4--dc23
LC record available at https://lccn.loc.gov/2021045035
LC ebook record available at https://lccn.loc.gov/2021045036



For more information, write to Bearport Publishing, 5357 Penn Avenue South, Minneapolis, MN 55419. Printed in the United States of America.

# CONTENTS

# THE INSIDE SCOOP

Your body is a super machine that keeps you moving, learning, and having fun. But how does it work? The secret is inside!

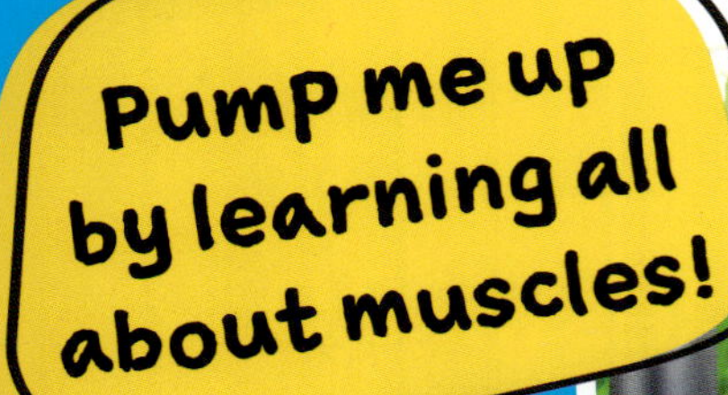

When you run, eat, or even blink, your muscles are hard at work. Without your muscles, your body would not be able to move. Let's take a closer look.

# MANY MUSCLES

There are around 640 muscles in your body. They come in many different shapes and sizes.

When you sit, you're using your biggest muscle. It's your bottom!

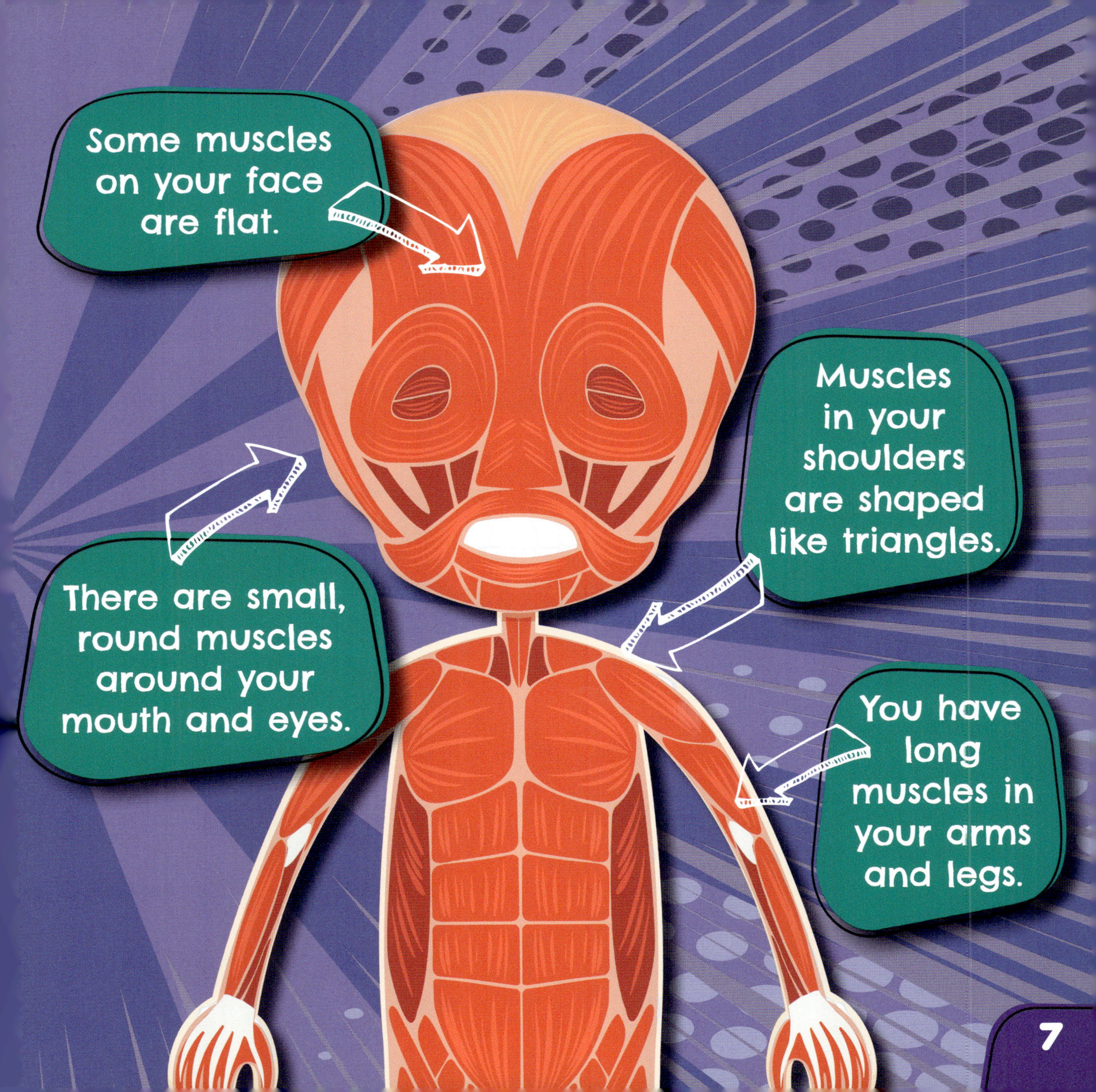
Some muscles on your face are flat.
There are small, round muscles around your mouth and eyes.
Muscles in your shoulders are shaped like triangles.
You have long muscles in your arms and legs.

# MOVIN' AND GROOVIN'

Muscles help your body move. You have three kinds of muscles.

Your heart is made up of **cardiac** (KAHR-dee-ak) muscle.
Muscles that connect to bones are called **skeletal** (SKE-lit-uhl) muscles.
I go by many names!
Smooth muscles work on their own. You don't control them.

# YOU MOVE ME

Skeletal muscles help you move the bones in your body. They join to your bones with **tendons**. When you want to kick a ball, you move the skeletal muscles in your leg.

There are no muscles in your fingers. They move with help from tendons.

Many skeletal muscles work in pairs. But as they work together, they may act as **opposites**. One muscle gets tighter while the other gets looser.

# OUT OF CONTROL

Your smooth muscles move even though you don't tell them to. They keep moving even when you are asleep!

Smooth muscles help your **lungs** breathe.
These muscles in your belly break down food.
They help you go to the bathroom.

# HEART WORK

Your heart is another muscle that moves all on its own. It's the hardest-working muscle in your body. And it never stops.

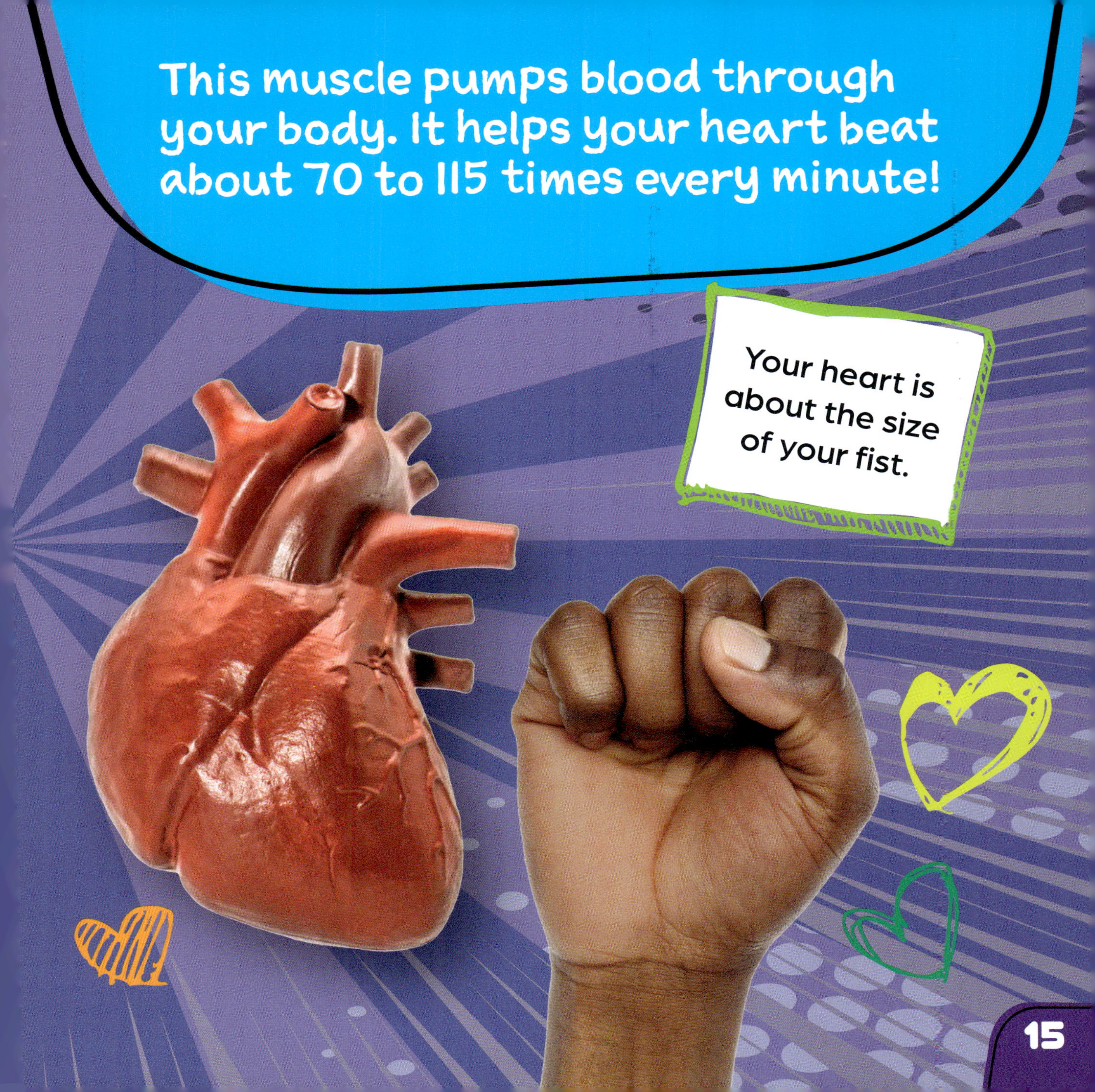
This muscle pumps blood through your body. It helps your heart beat about 70 to 115 times every minute!
Your heart is about the size of your fist.

# BRAIN POWER

How do your muscles know to move? Your brain tells them to! It uses **nerves** to send messages to the muscles.

It's time to get to work, muscles.

Whatever you say, brain!

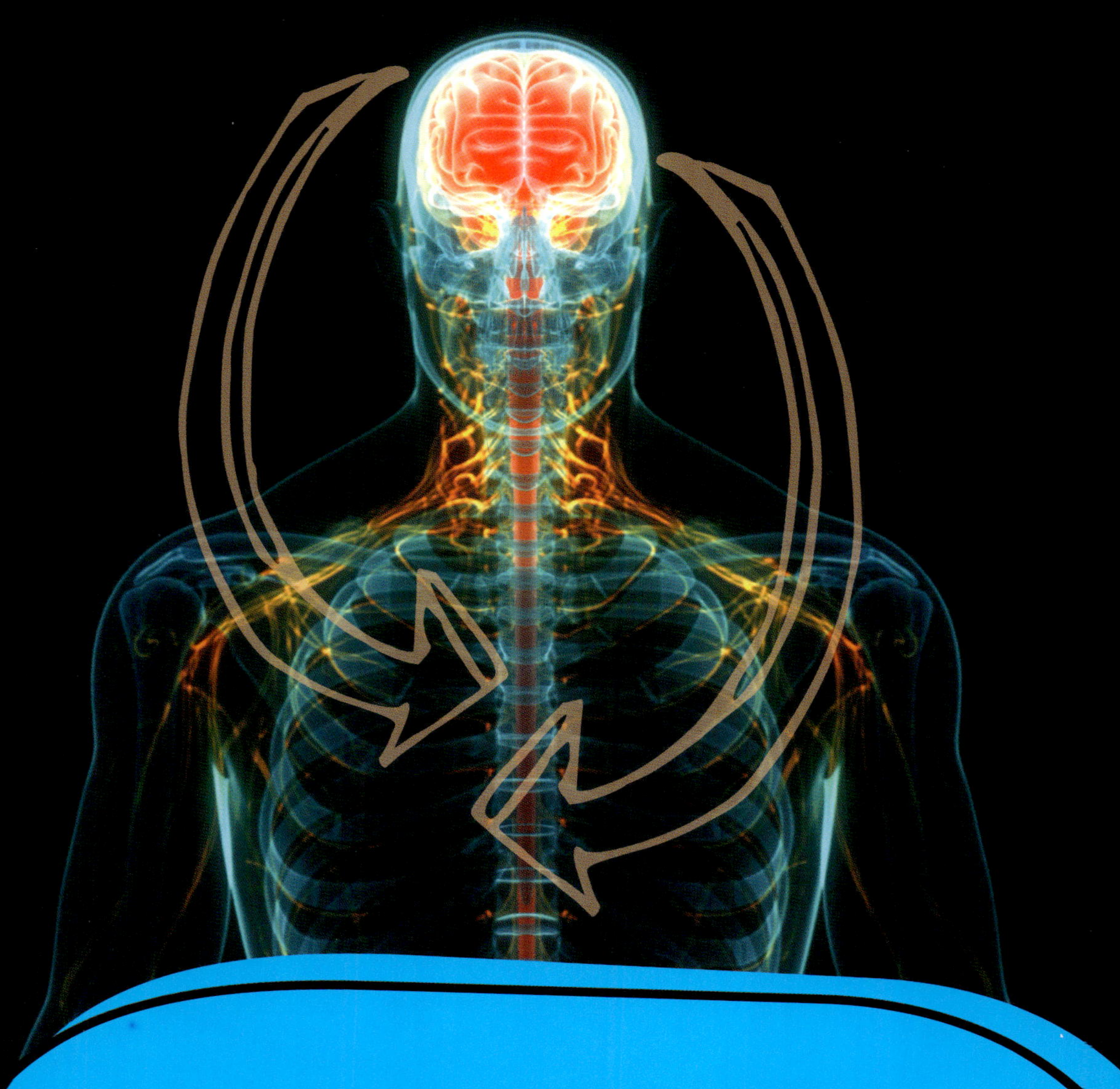

The left side of the brain controls the muscles on the right side of your body. And the right side of your brain is in charge of the left side!

# BIG AND STRONG

The bigger the muscle, the stronger it is. The only way muscles can get bigger is by exercise.

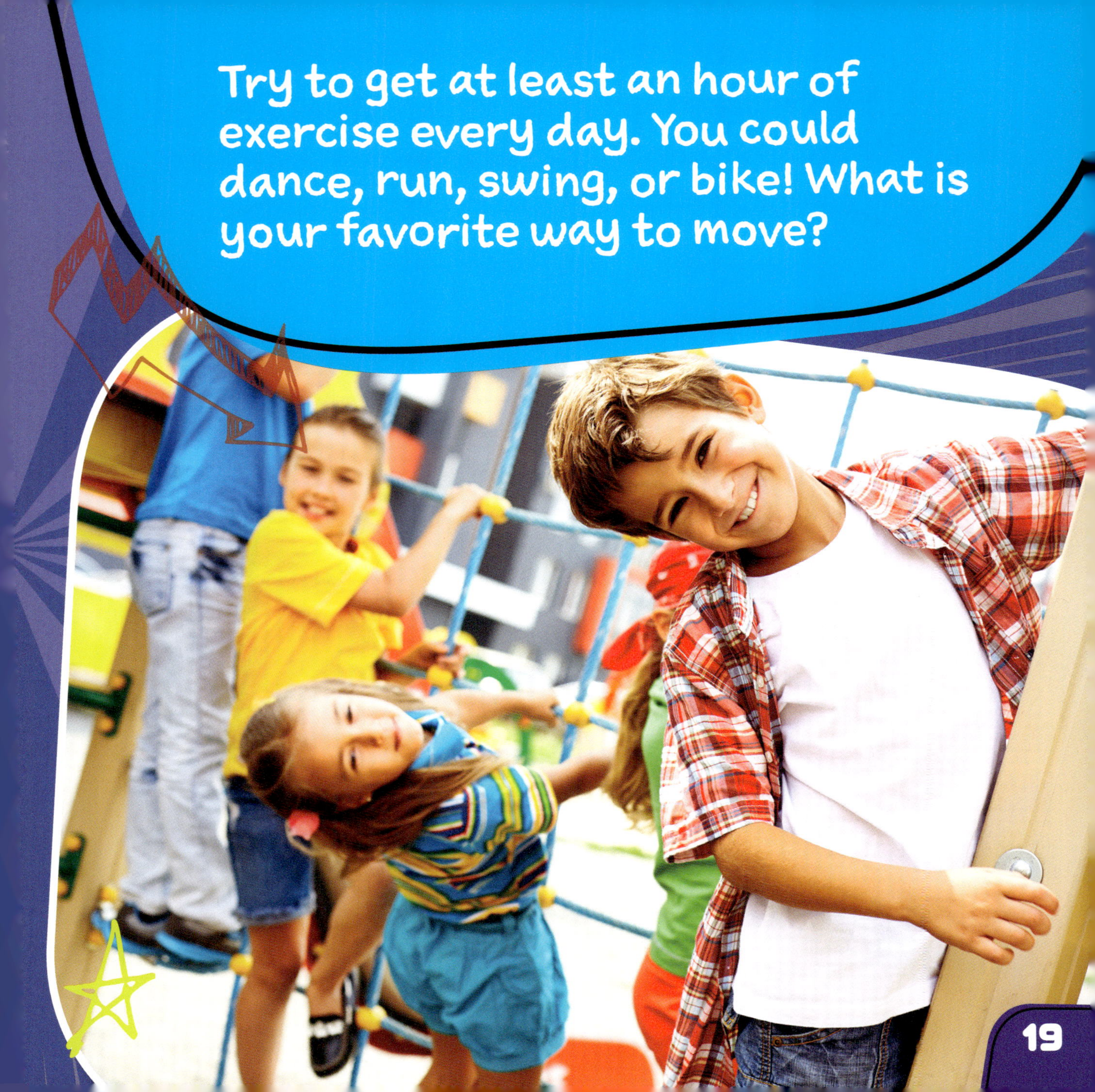

Try to get at least an hour of exercise every day. You could dance, run, swing, or bike! What is your favorite way to move?

# CARING FOR YOUR MUSCLES

Take care of your muscles so they can help you move for your whole long life. Before and after you exercise, **stretch** your muscles. This helps them warm up and cool down.

Simple choices you make every day will help your mighty muscles stay healthy for a long, long time.

# YOUR BUSY BODY

Your muscles are an important part of the super machine that is your body. They work with lots of other things inside you. Together, they keep you going every day!

It's what's inside that counts.

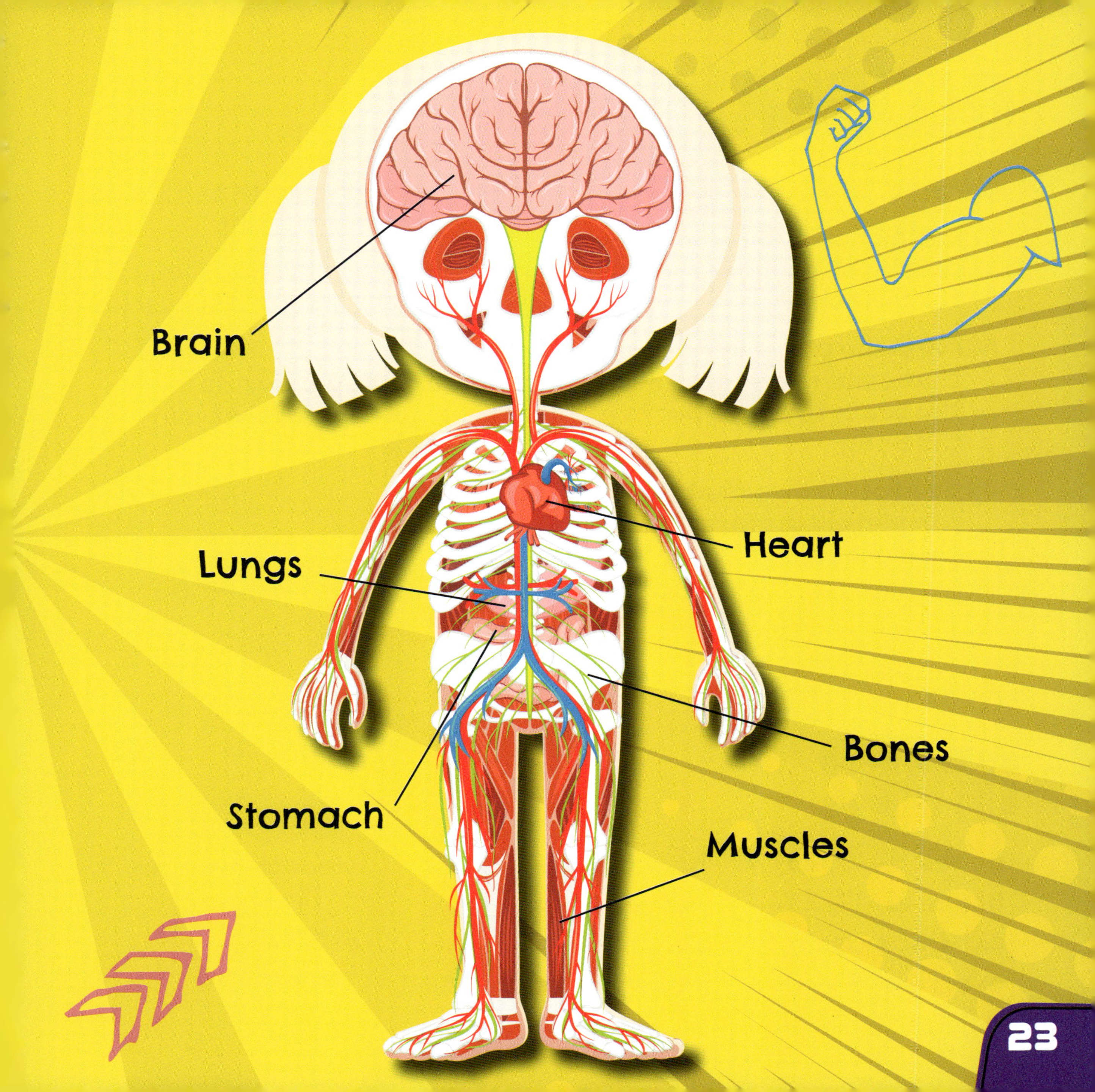
Brain
Heart
Lungs
Bones
Stomach
Muscles

# GLOSSARY

**cardiac** to do with the heart

**lungs** parts of the body in a person's chest that are used for breathing

**nerves** the parts of the body that send messages from the brain to other parts of the body

**opposites** things that are completely different from each other

**skeletal** to do with the bones in a person's body

**stretch** an exercise in which a person gently pulls muscles longer

**tendons** tough bands in a person's body that join muscles to bones

# INDEX